The Introduction

of

Queen Bees

by Brother Adams

Northern Bee Books

The Introduction of Queen Bees
© Brother Adams

ISBN 978-1-912271-16-0

Published by Northern Bee Books, 2018
Scout Bottom Farm
Mytholmroyd
Hebden Bridge HX7 5JS (UK)

Design and artwork by D&P Design and Print
Printed by Lightning Source UK

© Front Cover photograph by Erik Österlund.

The Introduction
of
Queen Bees

by Brother Adams

Forward by Colin Weightman

In March 1940, during the early months of World War Two, the renowned Somerset Beekeeper, Louis Edward Snelgrove, produced a Book entitled The INTRODUCTION OF QUEEN BEES.

Superbly printed by Purnell and set in bold type, with wonderful line drawings, they were much appreciated by older beekeepers. Snelgrove, examined and commented upon, all the well known methods of Queen Introduction then in use and gave beekeepers some original methods of his own, which he had developed - with his father - during many years of practical beekeeping. Indeed, the Snelgrove's were respected beekeepers away back into the last Century and they were highly thought of by George Jenner when he was CBI for Devon; a point confirmed by the former S.W. Regional Bee Inspector and noted practical beekeeper, Len Davie.

As a contemporary of the late Brother Adam, a friendly 'ding dong' went on between them on Beekeeping matters over the years. Even, with an Eighty colony outfit, of local bees (Apis melifera melifera) Bro. Adam always dismissed Snelgrove as being a small scale beekeeper. Bro. Adam's experiences of Introducing Queens were often at variance with the experiences of the Snelgrove's. He, was resolved, that, he too would publish something on the Introduction of Queen Bees and with this in mind he prepared ~ manuscript - which he circulated among his friends for comment - but, apart from providing material for his own lectures on the subject, nothing came of it. During one of his visits to my home, Brother Adam suggested that the material contained in the manuscript, might, one day, be of interest to a new generation of beekeepers: so, here it is.

I used to visit the Snelgrove's, at the Beeches, Bleadon, nr Weston on my way to Buckfast Abbey in the Nineteen Fifties and have fond memories of helping Louis Edward, and his daughter, with the bees. They, had developed their own strain of local dark bee and apart from the use of a smoker and hive tool, the docility of the bees was such, that no protective clothing was worn. Far, removed, from the days when Mr. Snelgrove's father had demonised the area with Cyprian crosses, a tale he recounted in his book on Queen Rearing (p.119) Snelgrove, constantly reminded Bro. Adam of his experience and of the DANGERS of cross breeding whenever they met.

Murton Park, York in 1989 – Festival of Food and Farming.

© Photograph by Mary Fisher.

INTRODUCTION OF QUEENS

The introduction of queens is without doubt one of the major problems of bee-keeping. Apart from the weather. which is beyond our control, the queen bee is the ultimate source and fountainhead of the well-being and productivity of a colony. By the mere substitution of a young vigorous queen we are enabled to renew the main-spring of the life of a colony - we have the power at all times to rejuvenate a colony and maintain it at its maximum productive ability; and, moreover, we possess the means to circumvent, to by-pass, most of the potential troubles with which the keeping of bees is beset.

Unfortunately, the introduction of queens as practised up to now results in many failures and difficulties. Indeed thousands of valuable queens perish annually on the very threshold of their useful existence because of a want of appreciation of the true cause which determines the acceptance of queens on introduction. By reason of an erroneous hypothesis, the underlying cause, the fundamental law at issue, which determines the acceptance of queens - though so simple and obvious - has remained shrouded in mystery and eluded the bee-keeper's grasp. False deductions have been drawn from the experience gained and the observations made, and the introduction of queens has proved an ever-more baffling operation with dire results.

Competent authorities assess the loss of queens in introducttion at 50% This estimate may be questioned by persons of limited experience. But judging from the losses we ourselves in former years sustained, the estimate of 50% does not, we fear, greatly exceed the number of queens that annually perish, or are rendered useless in introduction. The indirect loss arising from queens injured in introduction, often proves more grievous than the loss of queens that are killed outright. A colony headed by a defective queen, especially a queen which is not visibly injured, is of no real practical value: indeed such queens are more often than not a source of endless trouble.

We wish at the outset to emphasise and lay stress on this all important aspect of introduction. All too frequently queens are accepted, but injured to such an extend that they are accepted but injured to such an extend that they are superseded within a few weeks or months, and the bee-keeper is often totally unaware of the replacement silently effected by the bees. On the other hand, many queens are not visibly harmed, but they are nevertheless injured, and colonies in possession of such queens never prosper, never reach their maximum productive strength. Therefore it is of the utmost importance that no effort or means is ever neglected whereby it is possible to safeguard queens from sustaining any form of harm or injury in the process of introduction.

We regard this question of introduction of queens as one of the very few things that really matter in the management of bees. It is without any doubt the pivotal operation around which bee-keeping revolves. As far as we are concerned the problem of introduction of queens has been completely solved. The aim we set ourselves is twofold:- We must endeavour not merely to get every queen accepted, but also to get them every one of them - transferred to their permanent homes in full possession of their unimpaired vigour and fecundity.

Up to now all methods of introduction have been based on the supposition that before an alien queen is accepted she must somehow or other acquire the same odour as that possessed by the colony for which she is destined. It has been assumed that each colony possesses a distinctive odour. and that a new queen must by a period of confinement within a cage, in her new home, acquire an identical odour, or by some means or other be made acceptable must be "introduced" to the bees before they will take kindly to her. The question at once arises, is there any evidence which would warrant the assumption of the existence of individual colony odour ?

Recent investigations by competent scientists support our view that individual colony odour is not found characteristic of any colony.

No scientific proofs have so far been forthcoming in support of the commonly accepted assumption that every colony possesses its own distinctive odour, and that by it the bees are enabled to discern between members of their own community and those of other colonies. By the term "colony odour" it is assumed that bees emit a scent which imparts to every member of a colony a uniform and distinctive odour, and that this odour varies from colony to colony. No positive evidence however, has so far been found which would warrant this assumption.

There is a hive odour - a combination of odours derived from the combs, especially old brood-combs, propolis, honey, brood, etc. The hive odour undoubtedly varies in intensity and character, depending on the season temperature, honey-flow, etc. But the variation in intensity from colony to colony in the same apiary, subject to identical external conditions and only liable to minor differences inside the hive, can hardly form a sufficient distinguishing mark whereby the bees would be able to recognise one another. Indeed, as we shall presently show I there is definite evidence that hive odour does not in fact form an identification mark whereby bees recognise one another.

A queen bee undoubtedly possesses a particular odour, and bees are able to trace a queen by this odour. But does each queen emit a distinctive odour, and is this odour imparted to the bees?

Neither does the scent emitted by the scent gland, situated in the abdomen of the worker bees, seem to form any distinguishing mark. The purpose of this scent, when emitted by the bees, is to attract the members of a colony or swarm to some particular point.

But this scent is clearly not distinctive from colony to colony, otherwise no such obvious confusion would ever be possible as often happens when a number of swarms issue at one and the same time in an apiary and then return simultaneously. The attractive power of-this scent must be great indeed, for we know that any bees that are thus misled to join another swarm, or are thus enticed to enter a hive not their own will often be killed.

We have made a distinction between "Hive odour" and "colony odour" It is often implied that the former and the latter are one and the same thing. However, the following evidence will at least demonstrate that this is, in fact, not the case. The commonly accepted view that any pungent odour will obliterate colony odour is undoubtedly a mistaken notion. Oil of wintergreen, for instance, gives off a most penetrating odour, yet it has never been known to induce robbing. Indeed, methyl salicylate when used with reason seems to have no visible effect of any kind on bees. The Frow mixture, on the contrary, gives rise to robbing as nothing else will do, and robbing is the most serious objection to its use. Creosote will induce robbing. but in our estimation it is not the odour of the Frow mixture or of creosote that causes robbing, but we hold that the fumes of these substances induce a form of lethargy inhibiting the natural instinct of a colony from defending its stores. Though it is presumed that the Frow mixture and creosote mask colony odour, they definitely have no effect on hive odour. If they masked the hive odour, then assuredly no robbing would take place. On the other hand, why does Izal and carbolic acid repel robbers and the Frow mixture and creosote induce robbing ? If the assumption that any powerful odour ipso facto obliterates colony odour were the true explanation, then the Frow mixture, creosote, methyl salicylate , carbolic acid and Izal should in every case bring about identical results. However, having regard to the fact that hive odour is definitely not blotted out by the fumes of the Frow mixture or creosote, hive odour can therefore not be the distinguishing mark whereby bees recognise one another.

On the other hand, there is no definite evidence of the existence of such a thing as colony odour. Indeed, our observations and experience especially in regard to this matter the introduction of queens - would indicate that colony odour is a fictitious assumption. It is merely a hypothesis put forward as a convenient and plausible interpretation of phenomena and reactions on the part of the bees which have not yet yielded to any satisfactory explanation. We really do not know how bees are enabled to recognise one another. We have known a number of cases when, after the introduction of a queen by a certain method based on the idea that a queen must first aquire the colony odour so as to ensure acceptance, the most violent fighting ensued among the bees of the stocks to which the queens were introduced, so that the colonies were decimated leaving only a few bees and the queen alive after the fighting had ceased. No question of colony odour could arise in such a case. There may be such a thing as "colony odour", but all the evidence seems to disprove its existence.

Our experience has led us to the definite conclusion that "colony odour", even if there is such a thing, has nothing whatever to do with the acceptance of a queen. In every case, whatever the method of introduction employed, the factor which causes acceptance or rejection in introduction is primarily dependent on the behaviour of the queen. The behaviour of the Queen is in turn subject to her condition at the time she is liberated. Thus, for example, we are certain that the balling or rejection of a queen is due to her own behaviour. A newly mated queen or virgin when alarmed by the opening of the hive - even when the virgin or newly-mated queen has hatched in the same hive - will often get balled and killed. It frequently happens that when thus excited the virgin or newly-mated queen will rush about the combs and thereby rouse some antagonism amongst the rest of the inhabitants; and then is attacked at once. This is not only so when a hive is opened, it is liable to be caused by any undue excitement - notably in the case of virgin queens returning from a flight - and it is our opinion that whilst a few virgin queens may be killed by birds as they are flying in the air, or by entering into the wrong hives, a greater mortality is due to some cause of excitement within her own home, which engenders hostility amongst her own bees, and thereby brings about balling and the death of the queen. Balling in such cases cannot possibly be attributed to the absence of colony odour, for the virgin or newly-mated queen is part of the colony, and would of necessity possess the colony odour. But colony odour, if not a figment, has clearly nothing to do with the balling of a queen. We can lay it down as a principle: that the acceptance of a queen is determined by her condition and behaviour. We will explain what we mean by the "condition and behaviour".

If a young queen that has been laying for some weeks is caged and liberated in another colony the same day, within about twelve hours, she will be accepted without fail. If this same queen is not liberated until the second day, she is then liable to be attacked and balled. The reason is because the queen is not, or may not be, in a laying condition on the second day, whereas she is on the first. The longer a queen is confined the less likely it is that she will be accepted, unless the bees, while she is confined, feed her through the wire gauze of the cage and therefore when she is released she will immediately proceed with her normal activity of depositing eggs.

A queen sent by post and caged for three days in the hive for which she is destined will be accepted at the end of that time, provided the bees have begun to feed her through the wire gauze of the cage and thereby restored her ability to lay by the time she is released. If they do feed her in that way, and she has regained her laying condition, she is accepted at once; but if they do not feed her, and she is liberated before she is again in a laying condition and thus recovered from her imprisonment in the cage, she is attacked and killed; or if not killed she is balled and rendered useless.

For this reason a choice queen sent through the post should be always liberated in a nucleus, and the nucleus should be formed at least three days before the queen is due to arrive. By the time the queen is due, all the old bees will have returned to the parent stock, and, only the young bees are left in the nucleus. These will immediately feed the strange queen and thereby restore her vital functions and then accept her without must have attained full maturity, must be of a certain age, before she will be accepted whatever the mood of the bees to whom she is introduced. What we mean by the term "full maturity" as applied in this case, needs some elucidation.

A newly-mated queen, when she first commences to lay, is of a nervous and easily frightened disposition. The least undue disturbance under manipulation may endanger her life. However, in the course of a few weeks a profound change and transformation in her behaviour will gradually manifest itself. She will assume a more matronly gait and equanimity of behaviour, and will under manipulation calmly and sedately proceed with her normal routine. After laying for about four weeks, she will have attained full maturity. She will not be in her prime until the following year, but no further change in her behaviour will take place thereafter, excepting that with age her movements will grow progressively slower.

We have stipulated a minimum period of four weeks as the time required for a newly mated queen to attain her full maturity. We are aware however that some queens attain maturity somewhat earlier. On the other hand queens of an innate nervous disposition of mongrel, hybrid ,Native or French descent - may possibly take longer before they reach full maturity. But even in the case of queens of exceptionally nervous temperament after the lapse of two months, according to our experience, it will be safe to introduce them to any colony.

There is one further important aspect connected with this subject which demands consideration, namely, the serious injury caused to newly mated queens that are caged before they have attained full maturity.

The commercial queen-breeder endeavours to dispose of all newly-mated queens with the least possible delay, within a few days after they have commenced to lay. Apart from the grievous loss of queens in consequence thus sustained in introduction, we hold that newly-mated queens, if caged before they have reached full maturity, are exceedingly susceptible to injury, when imprisoned for any length of time at this stage of their development. We are firmly convinced that the failure of many valuable young queens must be attributed to this cause. A queen confined to a small cage for any considerable time, before she has attained full maturity, rarely if ever escapes injury.

The injury may not be noticed always, excepting by the most experienced bee-keeper.

A virgin or newly-mated queen is a most delicate creature, highly susceptible to permanent injury. Therefore, to secure the best possible queens - queens possessing the greatest possible stamina and fecundity the prudent bee-keeper will refrain from caging his queens at any stage of their development until they have reached full maturity.

Perhaps we ought to point out that, though we use the term "introduce", we do not in reality introduce queens in the ordinary accepted meaning of the word. We "substitute" a queen in place of another in every case. There is no process of introduction involved whereby the young queen is first made acceptable to the bees before they will, take kindly to her. A "substituted" queen, immediately on being liberated proceeds with her normal activities, regardless of her new surroundings - just as a bee returning from 'the fields laden with nectar, on entering another hive not her own, will proceed' with her normal duties as unconcerned as if she had entered her rightful home. A "substituted" queen, let it be repeated, is accepted as the rightful mother of a colony by virtue of her condition and behaviour.

Contrary to the view generally held, we maintain that a queen will never do as well in the season she is born as when she has reached her prime the following year. Moreover, a newly-mated queen when introduced at the height of the season to a large powerful stock will in succeeding years never do as well as the queens retained in nuclei until the autumn or the following spring. The idea that young queens are essential to wintering is a mistaken notion; and that young queens are indispensable for obtaining the largest possible surplus from the heather, is a misleading fallacy. According to our experience, a queen will furnish the most populous colonies for the moor, and the largest force of young bees for wintering when in her prime, that is the season following the year she is born. Therefore, requeening in July and August, for the purpuse of providing extra strong colonies of young bees for the heather honey harvest, or for wintering, is an utterly futile practice - except, of course, in the case of colonies whose queens are failing.

As a rule we never introduce a young queen to a honey producing stock the season she is born, except in years when we happen to have a surplus, of queens in the late autumn for disposal. We then introduce the surplus queens to the honey producing stocks during the first week in October, before they are finally closed down for the winter. But we prefer to leave the young queens in the nuclei, and then introduce them at the end of March, or as soon as the weather will permit in spring. We usually winter about four hundred nuclei. Not all these queens will be required for requeening in spring. The remainder are kept in reserve. If at any time later in the season we come across a colony in possession of a queen that is not up to standard, she is promptly replaced by one retained in reserve for this purpose.

We requeen our colonies mainly in the spring, and some of them occasionally in the late autumn, but merely for reasons of convenience.We change and introduce queens at all times of the season, whenever deemed desirable or necessary, irrespective of any other considerations and with infallible success. Indeed we now never trouble to ascertain whether a queen has been accepted.

However, we would sooner or later know if a queen had been rejected, for we never introduce a fertile queen without first clipping her wings.

We are not unmindful of the objections that may be raised by the practical bee-keeper against a scheme which demands the retention of young queens until the end of the season, or alternatively until the following spring. However, bearing in mind every aspect of the question, we are firmly convinced that requeening in the late autumn or early in spring, as advocated by us will inevitably in time supersede all other forms of annual requeening. Its advantages are so immense as to outweigh completely any drawbacks this method of requeening may possess. It must be remembered that:

1:     requeening is accomplished at the close or the beginning of the season, when no other important work demands the attention of the bee-keeper;

2:     at that time the change of queens is effected with the least possible disturbance or interference to a colony;

3:     the labour involved in giving new queens is reduced to the absolute minimum;

4:     there are no losses, and consequently fewer queens need be raised;

5:     every queen is accepted without sustaining any form of injury;

6:     the whole operation of requeening is reduced to an absolute certainty - in contradistinction to the notorious uncertainty, hazards, failures, and the untold consequential losses involved in the orthodox methods of introduction.

We have at one time or another put to the test, on an extensive scale, every well-known method of introduction of queens. We are, therefore fully aware of the advantages and limitations of each and every method. We suffered losses and paid the penalty. All the devices and schemes, every one of them, involve a greater or lesser element of uncertainty, and in a matter of such overriding importance the severely practical bee-keeper will, if at possible, leave nothing to chance. As far as we are concerned, we would not entertain the idea of using any of the orthodox means of introducing queens, any more than we would consider reverting to the keeping of bees in skeps.

We are nevertheless fully aware that at times circumstances may arise which necessitate the introduction of newly-mated queens before they have reached full maturity. Also, beekeepers who do not rear their own queens but rely on queens sent through the post, must resort to some other method of introduction and requeening than the one we have so far outlined and advocated. We shall, therefore, now deal with such cases and the like.

If for any reason it is desired to introduce a newly-mated queen to a queen-right or queenless honey-producing stock, before she has attained her full maturity, we know of no better and safer means than the nucleus method. Indeed, for a number of years, before 1937, we requeened all our colonies annually during the month of June and July by this method. It is not infallible, but the percentage of queens accepted is greater than that secured by any of the, ordinary methods of introduction in general use. Moreover, there is no need for caging the queen, and in consequence the danger of her sustaining any harm from imprisonment does not arise. As a matter of fact, we often even now resort to this manner of requeening whenever our reserve of one-year old queens is exhausted.

The nucleus method of requeening presupposes that the beekeeper rears his OWN queens and that the nuclei are situated in the same apiary as the stocks to be requeened. Or,alternatively that the queens purchased have been first introduced to newly formed nuclei and that in either case the newly-mated queens have been laying for at least a week.

The transference is effected as follows: the nucleus is first opened and the bees exposed to the light. Next, the queen of the colony destined to be replaced is searched for and deposed. Then three combs are removed from the brood-chamber of that colony. In their place is inserted, with as little disturbance as possible, the nucleus with the newly-mated queen surrounded by her own bees. The brood-chamber must then be left open exposed to the light far some five to ten minutes; then the supers are replaced and the hive closed. If desired one may the following day ascertain whether the substituted queen has been accepted, but it is wiser to wait a few days.

The combs of brood and bees which have been removed from the requeened hive are placed in the empty nucleus hive, and after the lapse of three days a queen cell inserted. Alternatively, if the nucleus is not required any longer, the combs and bees may be given to another needy colony.

The beginner will expect a fight to ensue when two lots of queen-right bees are thus united without any apparent precautionary measures being taken against such an eventuality. However bees exposed to the light for the duration of five minutes or so will peacably coalesce without any further safeguards. Here again in uniting of bees colony odour is of no significance. As far as we are able to interpret the facts success in uniting also depends on behaviour - in this case, of course, on the behaviour of the bees. The exposure to the light, as every observant bee-keeper will have noticed, has a calming effect on bees. We never at any time use any other precautionary measure but exposing to light to ensure an amicable union when equalising or when adding bees to a queen-right colony.

The careful reader will have noted that contrary to the usual procedure, we do not allow a period of queenlessness as generally recommended before the substitution of a new queen. In every case the old queen is removed and the young queen inserted immediately; excepting in the case of a queen sent through the post, which is given to a nucleus that has been newly formed a few days before the arrival of the queen. Our experience indicates that:

1:      there is no advantage of any kind derived from leaving an established colony queenless prior to the introduction of a new queen;

2:      when a queenless colony has begun to build queen-cells, the bees seem less friendly disposed towards a new queen though this applies only in cases where newly-mated queens have not 'reached full maturity;

3:      when queen-cells are found, everyone must be destroyed before the introduction of a queen, and if by any chance one is missed, though the new queen will be accepted she may subsequently be killed by the virgin, if the bees do not tear down the queen-cell.

A fully mature queen will be as readily accepted by a colony that has been queenless only a few days as one queenless for a number of weeks. The duration of queenlessness has no material effect on acceptance. But as already pointed out, whenever there are queen-cells present in the hive, there is a great probability of a cell being overlooked with fatal consequences to the new queen. Moreover. if a colony has been queenless over a long period, it may be in possession of a virgin queen. The surest means of ascertaining the presence of a virgin is by giving the colony a comb containing unsealed brood. If a virgin queen is present no queen-cells be begun.

We may now briefly summarise the points set forth. We have endeavoured, to show that:

1:     colony odour or hive odour does not possess any practical significance as regards introduction; 2:prolonged caging of queens more often than not renders acceptance more precarious;

3:     the condition of:the colony and the disposition of the bees have a bearing on acceptance, but that it is a practical impossibility to determine the psychological moment when a colony will infallibly accept a queen that has not reached full maturity

4:     the condition of a colony and the disposition of the bees is totally, immaterial - provided that the behaviour of the queen is in every instance such as not to arouse hostility;

5:     the behaviour of the queen is in every instance the deciding factor which ultimately determines acceptance or rejection;

6:     finally, the behaviour of a queen is dependent on her condition and age

The proofs we offer in support of our contention that the behaviour of the queen is the fundamental and only factor which determines acceptance are that:

1:      queens which have attained full maturity and that are in laying condition may be introduced in utter disregard of the prerequisites hitherto regarded as essential for safe acceptance;

2:      a colony thus requeened may, if desired, be opened the following day without the least danger to the new queen;

3:      acceptance of queens is infallible.

We are well aware that the foregoing statements are totally at variance with all the theories and teaching contained in our textbooks relative to this subject. But we are recording our experience and the facts as ascertained by us. We offer no apology for placing the utmost stress on this all-important matter of queen introduction. An infallible and safe method of introduction ensuring that every queen introduced shall not be merely accepted but accepted without sustaining any injury is a sine qua non of the bee-keeper. All competent authorities agree that the high percentage of young queens that annually perish on the very threshold of their useful existence, is of the most tragic and lamentable aspects of modern bee-keeping. There is truly no point on bee managemen which we consider more important than an infallible method of queen introduction. It is the key to success in bee-management. That, and the provisions of queens in sufficient numbers to supply the needs of the apiary.

Brother Adam 18th April 1992

www.ingramcontent.com/pod-product-compliance
Lightning Source LLC
Chambersburg PA
CBHW051433200326
41520CB00023B/7453